Caribou Migration

By Susan H. Gray

21st Century Junior Library

CHERRY LAKE PRESS

Published in the United States of America by Cherry Lake Publishing Group
Ann Arbor, Michigan
www.cherrylakepublishing.com

Reading Adviser: Marla Conn, MS, Ed., Literacy specialist, Read-Ability, Inc.
Content Adviser: Sheila K. Schueller, Ph.D., Lecturer and Academic Program Specialist, University of Michigan School for Environment & Sustainability
Photo Credits: ©longtaildog/Shutterstock.com, cover; ©Jukka Jantunen/Shutterstock.com, 4; ©Andrey Gontarev/Shutter stock.com, 6; ©nenets/Shutterstock.com, 8; ©RRichard29/Shutterstock.com, 10, 12; ©aaabbbccc/Shutterstock.com, 14; ©Petter W. Sele/Shutterstock.com, 16; ©Martin Capek/Shutterstock.com, 18; ©Franco Fratini/Shutterstock.com, 20

Cherry Lake Press is an imprint of Cherry Lake Publishing Group.

Library of Congress Cataloging-in-Publication Data

Names: Gray, Susan Heinrichs, author.
Title: Caribou migration / by Susan H. Gray.
Description: Ann Arbor, Michigan: Cherry Lake Publishing, [2021] | Series: Marvelous migrations | Includes index. | Audience: Grades 2-3
Identifiers: LCCN 2020002728 (print) | LCCN 2020002729 (ebook) | ISBN 9781534168558 (hardcover) | ISBN 9781534170230 (paperback) | ISBN 9781534172074 (pdf) | ISBN 9781534173910 (ebook)
Subjects: LCSH: Caribou–Migration–Juvenile literature.
Classification: LCC QL737.U55 G729 2021 (print) | LCC QL737.U55 (ebook) | DDC 599.65/81568–dc23
LC record available at https://lccn.loc.gov/2020002728
LC ebook record available at https://lccn.loc.gov/2020002729

Cherry Lake Publishing Group would like to acknowledge the work of the Partnership for 21st Century Learning, a Network of Battelle for Kids. Please visit http://www.battelleforkids.org/networks/p21 for more information.

Printed in the United States of America
Corporate Graphics

CONTENTS

5 On the Move

9 It's Still Cold!

15 Spring Migrations

19 Another Trip

22 Glossary

23 Find Out More

24 Index

24 About the Author

Caribou can weigh up to 700 pounds (318 kg).

On the Move

A sudden cold snap brings on the first snowstorm. The weather is changing. And caribou herds all across northern Europe, Russia, and North America are sensing it.

Young and old caribou have been chomping on moss, grass, and mushrooms all summer long. They have put on weight for the trip ahead. Even the calves have gained more than 100 pounds (45 kilograms)! Like

Caribou have the longest migration of any land animal.

their parents, they have prepared to **migrate** south.

Caribou live in herds. A herd may include a few thousand animals. It can have as many as half a million. Dozens of herds exist across the caribou's **range**. Every year, herds migrate back and forth between their summer and winter homes. Herds do not all travel the same distances. Some plod barely more than 100 miles (161 kilometers). Others migrate more than 1,500 miles (2,414 km) each way.

Caribou have four "toes" on each foot, which help them not slip on ice.

It's Still Cold!

The change in weather tells the caribou something. They cannot stay in their summer homes much longer. The weather will grow bitterly cold and windy. Food will become difficult to find. So around August, the herds start to migrate south. They head to areas where the winter blasts are not as awful. But it will still be cold. Snow covers the ground and rivers are icy.

When their path is blocked, caribou have to cross icy rivers.

Caribou are built for winter weather. During their migrations, they may face cold, deep rivers. But caribou are great swimmers, thanks in part to their wide feet.

Each animal's coat has two layers. The outer coat has long hairs. These hairs are hollow, with air trapped inside. The trapped air helps to keep caribou afloat. The other

Think!

Caribou won't cross rivers that are fast and wide. They look for better places to cross. If they are such good swimmers, why do they care?

One male caribou will mate with several females.
Males fight to prove their worth.

layer is a thick, furry undercoat that provides warmth.

As they move south, caribou continue to feed and gain weight. The herds reach their winter grounds around October. Soon, males and females mate. But baby caribou won't appear until next May.

Look!

Caribou use their front feet to scrape away snow that blankets their **forage**. Then, they nibble on their food. Look at pictures showing caribou noses and lips. What protects them from getting frostbite?

Caribou have a good sense of smell. They can detect plants and **lichens** under the snow.

Spring Migrations

In March, caribou begin moving back north. They are going to their summer **calving grounds**. They do not go in a straight line. Instead, they wander this way and that.

All winter, they scraped at the snow. They gnawed what little was underneath. They are hungry! Now, green shoots are finally pushing through the soil. They roam, eating whatever they can.

Calves can stand and walk on their own the same day they are born.

Caribou reach their summer homes around April. In May, the females have their calves within a few days of each other. Soon, each new baby is trotting around with its mother. Such athletic calves are good at escaping **predators**. The herd also provides protection for them. In the herd, thousands of eyes are watching for danger.

Ask Questions!

Baby animals that are up and moving about quickly are called **precocial**. Look online for other precocial animals. What are some advantages they all share?

Both male and female caribou grow antlers.
Each season, their antlers fall off and grow back.

Another Trip

At the calving grounds, young caribou run and play. There's plenty of food. Things are pretty good. That is, until the bugs come. First, it's the mosquitoes—millions of them. To get away, caribou keep moving about or seek windy areas. Then, it's the warble flies and the nose botflies. Warble flies lay eggs on caribou skin. Nose botflies place **larvae** in their noses. Caribou shake their heads and run like crazy to get away from these pests.

Climate change is impacting caribou migration patterns.

Finally, the insects leave and the caribou get some peace. A few weeks after the flies depart, the animals notice something. Cold weather is returning. The first few snowflakes are falling. Green plants are starting to shrivel. Caribou begin to look south. It's time again to leave their summer home.

Make a Guess!

Do you think that nose botflies bother any other animals? If so, which ones? Look online to see if you are correct.

GLOSSARY

calving grounds (KAV-ing GROUNDZ) areas where animals go to have their calves

forage (FOR-ij) food for cattle, horses, deer, and similar animals

larvae (LAR-vee) very young, undeveloped insects

lichens (LYE-kuhnz) plantlike organisms composed of fungi and algae that grow on rocks and trees

migrate (MYE-grayt) to move from one region to another and back again

precocial (pree-KOH-shuhl) very active and moving about shortly after birth or hatching

predators (PRED-uh-turz) animals that hunt and eat other animals

range (RAYNJ) the normal region where an animal exists

FIND OUT MORE

BOOKS

Dinmont, Kerry. *It's a Caribou!* Minneapolis, MN: Lerner Publishing, 2019.

Jaycox, Jaclyn. *Caribou Are Awesome*. North Mankato, MN: Pebble, 2020.

Patrick, Roman. *Caribou*. New York, NY: Gareth Stevens Leveled Readers, 2010.

WEBSITES

BioKids—Caribou, Reindeer

http://www.biokids.umich.edu/critters/Rangifer_tarandus
See this site for facts about caribou's lives, their antlers, males clashing, and the differences between caribou and reindeer.

National Geographic Kids—Caribou

https://kids.nationalgeographic.com/animals/mammals/caribou
Check this site for terrific photos, more information about caribou, and a great range map.

INDEX

antlers, 18
August, 9

calves, 5, 13, 15, 16, 17, 19
caribou
  range, 7
  weight, 4, 5, 13
climate change, 20
coats, 11, 13

Europe, 5

feet, 8, 11, 13
females, 12, 13, 17, 18
food, 5, 9, 13, 14, 15, 19

herds, 5, 7, 9, 13, 17

insects, 19, 21

males, 12, 13, 18
mating, 12, 13
migration, 6
  changing patterns of, 20
  how far caribou travel, 7
  southern, 9–13
  spring, 15–17

North America, 5

pests, 19, 21
predators, 17

rivers, 9, 10, 11
Russia, 5

smell, sense of, 14
summer, 5, 7, 9, 15, 17, 21
swimming, 11

weight, 4, 5, 13
winter, 7, 9, 11, 13, 15

ABOUT THE AUTHOR

Susan H. Gray has a master's degree in zoology. She has written more than 170 reference books for children and especially loves writing about animals. Susan lives in Cabot, Arkansas, with her husband, Michael, and many pets.